AF600908

German Aircraft in Hungarian Service

Zoltán Czirók

Great War Aviation Centennial Series #92

Notes

For our aviation books, please see our website at: www.aeronautbooks.com.

I am looking for photographs of the less well-known German aircraft of WWI to complete this series. For questions or to help with photographs you may contact me at jherris@me.com.

Interested in WWI aviation? Join The League of WWI Aviation Historians (**www.overthefront.com**), Cross & Cockade International (**www.crossandcockade.com**), and Das Propellerblatt (**www.propellerblatt.de**).

ISBN: 978-1-964637-08-2

Cover Painting: Tomasz Sadzinica
Color Profiles: Alexsandr Kazakov
Design and layout: Jack Herris
Cover design: Aaron Weaver
Digital photo editing: Jack Herris

Aeronaut Books
Books for Enthusiasts by Enthusiasts
www.aeronautbooks.com

Table of Contents

Preface and Acknowledgements

German-built aircraft played an essential role in the independent Hungarian air arms established after World War I, even though Hungarian aircraft factories continued their production after November 1918. The Hungarian Soviet Republic was declared on 21st March 1919 and immediately entered a state of war. Its air troops needed an increasing number of high-quality, high-performance aircraft, which they had access to from two sources.

The German Fokker D.VI and D.VII fighters were assembled at MÁG in Matyásföld, and most of them served with the Hungarian Red Flying Troops; this book reveals hitherto unexplored details about their fate. However, another source of German aircraft takes the reader into an even less known area: the squadrons of the German Mackensen Army fighting on the Balkan front, about which only limited sources are available, especially from the last period of the war. Hungarian archival documents have now made it possible to identify most of the aircraft of these German squadrons, the details of their later deployment, and their fate, with the help of sources discovered by Czech, Romanian, and Serbian researchers. Hopefully, this information will also advance further research into the air war on the Salonika front during World War I.

This is the first time to give the never-before-told detailed history of WWI German warplanes in postwar Hungarian service in the turbulent period when Hungary had to fight for its borders with Czechoslovakia, Romania, and the Kingdom of SHS, which required a lot of assistance. The author would like to warmly thank the following aviation historians for their numerous information exchanges, documents, and photographs, which greatly enriched this book. Without them, this book could not have been written: Christian Reiterer (Austria), Jan Kaše and Petr Lukeš (Czech Republic), Thorsten Pietsch and Reinhard Zankl (Germany), Zoltán Csizek and László Winkler (Hungary), Dan Antoniu and Horia Stoica (Romania), Boris Ciglić (Serbia).

Heritage from the Austro-Hungarian Era

Fokker D.VI

The Fokker V 12 rotary-engined biplane prototype (w/n 1980) was shipped to MÁG at Mátyásföld on 3rd January 1918 without engine and armament. After installation with a 150 hp Le Rhône (St) rotary engine at MÁG, the aircraft took part in the Second Fighter Competition at Adlershof in early May 1918 along with two other new MÁG-Fokker prototype fighters. Demonstrating the aerobatic qualities of the plane by Anthony Fokker himself, the LFT purchased the prototype and 14 Fokker D.VI fighters in June 1918 due to favorable impressions. The V 12 flew also at the July Fighter Evaluation.

The first seven fighters of the ordered production machines were shipped to MÁG on 27th August 1918 for engine and armament installation (for 23.500 Marks/piece). Due to the delivery delay of the total number of D.VI aircraft, a contract modification has already been considered earlier. On 6th August 1918, the Fokker aircraft company informed the *Luftfahrguppe* in Berlin that "the remaining 8 aircraft are to be delivered as Fokker monoplanes type E.V for 160 hp Le Rhone or Steyr engine" based on a personal consultation between Friedrich W. Seekatz and *Generalmajor* Emil Uzelac, head of the *k.u.k. Luftfahrtruppen*. A month later, however, the Austro-Hungarian War Ministry requested to obtain the delivery of eight Fokker D.VII type aircraft instead of the D.VIs (and also the E.Vs) because of the current maintenance difficulties of the Steyr rotary engines.

Austro-Hungarian Fokker D.VI

Word Numbers	Serial Numbers	A-H Designation*
2614	1632/18	04.101
2615	1633/18	04.102
2616	1634/18	04.103
2617	1635/18	04.104
2623	1641/18	04.105
2624	1642/18	04.106
2628	1644/18	04.107

*assuming that designation numbers were applied in sequence

Above: MÁG director Anthony Fokker posing in front of the Fokker V 12 (w/n 1980) fighter at Mátyásföld in the spring of 1918. (Miklós Liptai collection via Zoltán Czizek)

Fliegerarsenal (Flars) sent five 150 hp Le Rhône rotary engines built by the Waffenfabrik Steyr (07017, 07020, 07021, 17023, 07025) to Mátyásföld for the Fokker D.VI aircraft already with the order dated on 26th July 1918, and engine installation at MÁG was nearing completion in late 1918. The D.VI fighters were assigned the LFT designation 04.101 to 04.107.

The end of the war found the planes were at MÁG. On 14th November 1918, the seven Fokker D.VI fighters were planned to be assigned without machine guns to *1. légi rendőrségi osztály* (1st Air Police Flight) at Mátyásföld, one of the first units of the newly established Hungarian

Left: Test pilot *Hauptmann* Oszkár Fekete standing by the Fokker V 12 fighter. (Miklós Liptai collection via Zoltán Czizek)

Flying Troops. However, there is only information about one D.VI fighter being operational at that time. On 16th November, the Hungarian People's Republic proclamation, all available aircraft around Budapest were ordered to emerge over the capital and surrounding cities to shower the celebrating crowds with leaflets and flowers. The unit at Mátyásföld took part in the ceremony with four one-seater aircraft, including Fokker D.VI 04.106, piloted by *Oberleutnant* Antal Lányi. The planes took off at 11.50 a.m. and flew over the Parliament, dropping the leaflets and landing at the airfield at 12.20 p.m. After that, the fate of the planes developed differently.

The famous engineer *Leutnant* Vilém Žurovec, inventor of the PKZ 2 helicopter, worked on his helicopter project in Budapest with his brothers, Josef and Leopold, when the war ended. Josef and Vilém brought two Fokker D.VI aircraft to their home village, Harty (it was demolished and razed in 1956 during the

Above: LFT and MÁG personnel in front of the Fokker V 12 at Mátyásföld. (Miklós Liptai collection via Zoltán Czizek)

Below & Following Page: Fokker V 12 at Aspern. The question mark is a sign that the designation was not known to the airfield mechanincs. (László Winkler)

construction of the Ostrava-Mošnov Airport; its former territory is now part of Petřvald, Czech Republik), and they also fetched nine aircraft engines of various types, both rotary and straight, in 1919. Josef rebuilt one of the D.VI planes as a two-seater, and the second was sold to Prague. Josef Žurovec flew the two-seater D.VI until he crashed with it in 1921.

Three Fokker D.VI fighters are known to have entered into service with the Hungarian Flying Troops. Their number first appeared on a list dated 3rd February 1919: 04.101 (07021), 04.102 (07010), and 04.103 (07025) were attached to *1. repülőosztály* at Mátyásföld. On 12th March, the planes were not operational due to an oil shortage, and machine guns were still not installed on 28th April 1919. In the meantime (from 8th April), the name of this unit was changed to *8. harci repülőszázad* and was exclusively equipped with single-seat aircraft, the only field unit of its kind in the Hungarian Red Army.

Unfortunately, only a few records of the deployment of Hungarian D.VIs remained during the fights of the Hungarian Red Flying Troops. However, the one or two Fokker D.VIs were flown primarily by pilots Viktor Storer and Sándor Kasza, who, along with other routine squadron members, were constantly involved in the unit's sorties. One such specific incident is known: on 29th July, Hungarian planes took off to intercept an enemy aircraft. As the report states, a French three-engined (?) aircraft was spotted flying from Debrecen in the route of Onga–Miskolc towards Izsó at 10.15 a.m. Three Fokker D.VIIs (with pilots László Újvári, János Risztics, and Sándor Kasza) and one Fokker D.VI (Viktor Storer) of *8. harci repülőszázad* went in pursuit of the aircraft but failed to intercept it.

After the breakthrough of the Hungarian lines at the Tisza River by the Romanian troops and the subsequent collapse of the Hungarian Soviet Republic on 1st August 1919, the Hungarian squadrons that had the opportunity to do so retreated to the hinterland, to the western border of the country. *8. harci repülőszázad* was first withdrawn to Győrszemere, but after a short time, they moved to Szombathely with *4. repülőszázad* which was stationed primarily at Győr. After the Romanian occupation of Budapest, the Romanians began a systematic raid for materials, and of course, the aircraft factories were no exception. However, there was an error in the account of the aviation material carried away by the Romanians, as the Hungarian summary indicates that six Fokker D.VIs with engines, two without engines, and one experimental D.VI were taken from MÁG – a larger quantity than

Above & Below: *Oberleutnant* Antal Lányi in flying clothes with his Fokker D.VI fighter at Mátyásföld in the late autumn of 1918. The photos were published in a contemporary Hungarian aviation newspaper. (Aviatika)

Above: Fokker D.VI 04.102 in the service of the Hungarian Flying Troops. (László Winkler)

the number of aircraft that were shipped to the factory, not to say the planes of *8. harci repülőszázad*. Romanian documents mention two Fokker D.VIs captured in Hungary in 1919: number "16420" and "4105", which are probably 1642/18 (supposedly 04.106) and 04.105, while photos show 04.105 and 04.107 in Romanian service. According to the limited available sources, two D.VIs were assigned to the Grupul 5 Aviație (based on the airfield at Szamosfalva/Someșeni near Kolozsvár/ Cluj-Napoca from 20th August 1920) and "16420" was under overhaul in 1922.

In Hungary, most of the aircraft and personnel had retreated to Szombathely were assigned to the newly-formed local squadron, which was set up within the so-called National Army and led by the former commander of *8. harci repülőszázad*, László Háry. At the beginning of November 1919, six planes were flown from Szombathely to Budapest, preparing for the new governor, Miklós Horthy's entry into the capital on 16th November 1919, including a Fokker D.VI fighter. There is no further information on the deployment of this or any other D.VI, but apparently, they were hidden from the Inter-Allied Air Control Committee in 1921.

When aviation in Hungary slowly resumed after the end of the prohibition on flying and aircraft construction, the development of the aircraft repair plant, which had previously been relocated from Szeged, was started in Székesfehérvár. At Székesfehérvár-Sóstó airfield, under the leadership of engineer Elek Fabinyi, the Central Repair Plant was initially engaged in the repair of hidden aircraft and, among others, a Fokker D.VI was handed over to the air force in 1926. This last remaining D.VI was withdrawn from service and was written off in the same year, thus ending a career of this type in Hungary.

Above: Hungarian Fokker D.VI of *8. repülőszázad* in the air, propbably with Sándor Kasza in the pilot's seat in the summer 1919. (László Winkler)

Right: An unidentified pilot of *8. repülőszázad* posing in front of 04.101 on the provisional airfield at Mezőtárkány on 5th May 1919. (István Kern)

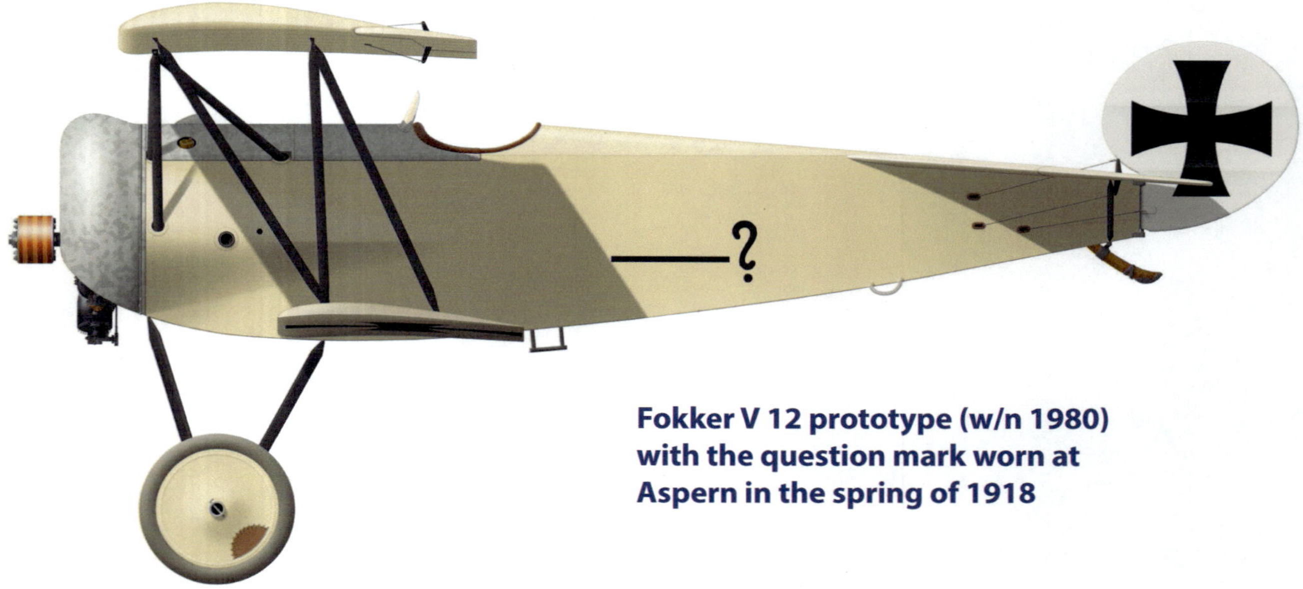

Fokker V 12 prototype (w/n 1980) with the question mark worn at Aspern in the spring of 1918

Fokker D.VI 04.101 of the Hungarian Red Flying Troops. The plane served at *8. repülőszázad,* the only Hungarian fighter squadron in the era of the Hungarian Soviet Republic.

Fokker D.VI 04.105 with Hungarian markings

Fokker D.VI 04.105 with Romanian markings after its capture

Above: The two-seater Fokker D.VI that Josef Žurovec rebuilt. (Jan Kaše)

Right: Fokker D.VI 04.105 in Romanian hands. The repainted place of the Hungarian markings of the red star on the fuselage and the upper wing is visible. (Horia Stoica)

Left: Another photo of a Fokker D.VI in Romanian colors. (Horia Stoica)

Fokker D.VII

Above & Following Page, Above: Anthony Fokker after a demonstrating flight with the V 22 at Mátyásföld in May 1918. In the background is the Fokker 90.03 (V 7) triplane and Aviatik D.I (MÁG) 92.14. (Miklós Liptai collection via Zoltán Czizek)

After the Fokker V 22 (w/n 2342), which had a standard D.VII steel-tube fuselage and powered by a 200 hp Daimler engine, was tested on 15-16 March 1918 at Schwerin, the aircraft was shipped to Matyásföld on 24th April 1918. Despite its excellent performance and maneuvrability highly praised by the Flars test pilot *Stabsfeldwebel* Franz Kuntner, the LFT refused the production of Fokker D.VII. Anthony Fokker demonstrated the capabilities of the V 22 for Flars engineers at Mátyásföld in May 1918, and the aircraft designated as MÁG-Fokker 90.05 was also flown at the Fighter Evaluation on 9-13 July. However, its most impressive flights were presented at Aspern on the 25th and 26th of July, which convinced *Generalmajor* Uzelac. As a result of negotiations in August 1918, the production of 660 Fokker D.VII fighters was decided, with delivering between December 1918 and March 1919 and built by four Austro-Hungarian factories (Aviatik, Fokker, MÁG, and Thöne & Fiala).

Flars ordered 75 D.VII fighters without engine and armament from Fokker in August 1918, and the first ten fighters were scheduled to deliver in October 1918. Since the 225 hp Daimler (MÁG) engine was in production at MÁG then, this engine was chosen for installation in the Fokker D.VIIs. The first six airframes (w/n 3861, 3863 to 3867; price: 23.500 Marks apiece) were shipped to MÁG on 23rd October 1918, which finally reached Mátyásföld despite delaying in the last days of the war.

According to the aircraft register of the Hungarian Flying Troops, seven Fokker D.VII fighters were at Mátyásföld on 14th November 1918 "with machine guns, offered for sale". The seventh plane was probably the first MÁG-built D.VII (93.01), as these fighters were listed in the inventory of the *1. repülőosztály* at Mátyásföld on 12th March 1919. All the fighters of German origin were marked as "under assembly at the factory. Not operational. The original German D.VII fuselages will be fitted with the 23000 series MÁG-Daimler". Nevertheless, the MÁG-built Daimler engines were also listed as a part of the aircraft: 3861 (23455), 3863 (23465), 3864 (23466), 3865 (23473), 3866 (23474), 3867 (23476). On 28th April, all six German D.VIIs were ready to fly and "now being machine-gun mounted".

The fate of only one of these machines is known. The reorganization of Hungarian Red Flying Troops on 8th May 1919 did not define specific tasks for the *8. harci*

Above & Facing Page: Poor quality but rare photos of the Fokker-built D.VII fighters of the Hungarian squadron at Mátyásföld. (László Winkler)

Above: Rákos airfield in the late summer of 1919. On the right side, three Fokker-built D.VII fighters are visible. (László Winkler)

repülőszázad. Still, a much-discussed event revealed that a detachment of the unit had been redirected to support the fighting around Salgótarján. On 13th May, István Fejes made an emergency landing in his German Fokker D.VII fighter No. 3867 at Pinc (now Pinciná, Slovakia) near Losonc (now Lučenec, Slovakia), and the pilot was captured after hiding for half a day, between 9 a.m. and 4 p.m. as a contemporary Hungarian newspaper article stated. According to Fejes' testimony, after his capture, he received orders to relocate to Salgótarján on 12th May, where he arrived the same day with two other pilots. In this light, the report of 12th May about László Újvári, István Fejes, and Nándor Udvardy landed their Fokker D.VII fighters at 10.45 a.m. in Zagyvapálfalva – i.e., they moved a detachment of three aircraft to the makeshift airfield at the border of Salgótarján – makes sense. On the next day's mission, Fejes' task was to determine whether Czechoslovak troops had received reinforcements or had left the town as a Hungarian attack was preparing on Losonc from two sides. Fejes stated in his account after his release that he had been attacking an armored train at Losonc when the synchronizing device of his machine guns failed and shot through the propeller, forcing him to land. There is a version of events in Czechoslovak historiography in which Fejes was shot down by a Czechoslovak reconnaissance plane, but this is disputed by Czech aviation historians, too. The "victorious" pilot was *nadporučík* Josef Pospíšil, commander of the *2. polní letecká setnina* (2nd field squadron), who allegedly forced Fejes to land. The source is a document written after the event, probably in conjunction with details of other events. In contrast, any official military document of *nadporučík* Pospíšil does not mention the air victory, and he did not take off that day, either, like none of his squadron's aircraft. The fact that the landing site was outside of the operation area of this Czechoslovak unit is a further argument against this, as well as the victory of a two-seater Hansa-Brandenburg C.I against a superior Fokker D.VII with an ace of 16 confirmed victories in the pilot's seat is also highly improbable. At the same time, the propeller of the crashed Fokker does not show any significant damage in the contemporary photos, which also calls Fejes' account into question. In any case, the Hungarian fighter plane fell into enemy hands, essentially unharmed.

After the emergency landing, Fokker D.VII 3867/18 was sent for overhaul to Prague on 13th May, and after a sortie on 4th July by pilot Rudolf Valenta, it

Above: István Fejes' Fokker D.VII 3867 in Czechoslovakian hands, on a rail waggon. (Petr Lukeš)

was used for propaganda flying. It is not known when the Schwarzlose machine guns were dismantled, but the last mention of the armed D.VII is in the records of 15th August 1919 and the photograph of 7th September 1919. At the beginning of 1920, it served in the pilot school in Prague, and on 1st June, it got a new engine of the same type, the Austro-Daimler Nr. 23266 coming from the Albatros D.III (Oef) 253.116 crashed by Adolf Bláha on 3rd November 1919. In October, it was flown to the Aircraft Training Center at Cheb, and on 2nd December, due to the considerable popularity of the aircraft, a request was made to arm it with Vickers machine guns. From the end of 1919, designation 38.67 was painted on the fuselage, an adaption of its original work number.

On 24th January 1921, Augustin Charvát flew the plane to HLD (*Hlavní letecké dílny* = Main Aviation Schops) in Prague for minor repairs. Still, the aircraft underwent a complete overhaul, including the engine, and was finally armed with Vickers machine guns. The Fokker was handed over on 21st April after the repair, but the engine problems were over in July because, on 5th July 1921, the repair was requested from HLD with the promise that it would be ready in a month. 38.67 got into the air only on 7th September after the repair. Then, on 9th September, he participated in the 1st air meeting arranged by the Czechoslovakian Aviation Club in Prague (with number "16" and pilot *nadporučík* Jan Popelák) and returned to Cheb on 27th September. During the flight from Cheb to Prague on 18th October, pilot František Procházka had engine problems that culminated in a crash after his ejection in Bohnice. The wreckage of the Fokker was stored.

Fokker 38.67 was demolished in Cheb on 15th January, but a day later, its fuselage was carried by train to HLD in Olomouc, where Josef Pospísil successfully flew it on 3rd June after a year of repairs. For the rest of the year, the aircraft continued to serve without any significant events in Cheb, but after the repair, it no longer carried machine guns and used Sigma and Knoller-Járay propellers. After the correction on 22nd January 1923, only a photograph is known of it from Olomouc on 29th July 1923, but it still belonged to the Training Center at Cheb. It underwent further repair on 23rd January 1924 at HLD Olomouc, and a day later, it was ordered to fly and continue to serve at Cheb. Still, nothing is known of its activity until 8th July, when it was found suitable for the school circuits in the one and two-seater pilot school. By the order dated 15th September

Above & Facing Page: Fokker D.VII 38.67 with two- and four-bladed propellers and different national markings. (Petr Lukeš & Miklós Liptai collection via Zoltán Czizek)

1924, it was to be scrapped, therefore, on 14th October, it was handed over by the Cheb school to the warehouse. It was decommissioned on 15th January 1925. The only Czechoslovakian Fokker D.VII flew 33 hours and 57 minutes of flight during the whole time.

German Fokker D.VII 5146/18 captured by Hungarians coming from the Balkans with the marking of the skull and crossbones

German-built Fokker D.VII w/n 3867 with the original four-color lozenge camouflage and the Hungarian red star in May 1919

German-built Fokker D.VII w/n 3867 with the early Czechoslovakian national markings after its capture. The plane served in this form from 1919 until October 1921, and the number 38.67 was painted on the fuselage at the end of 1919

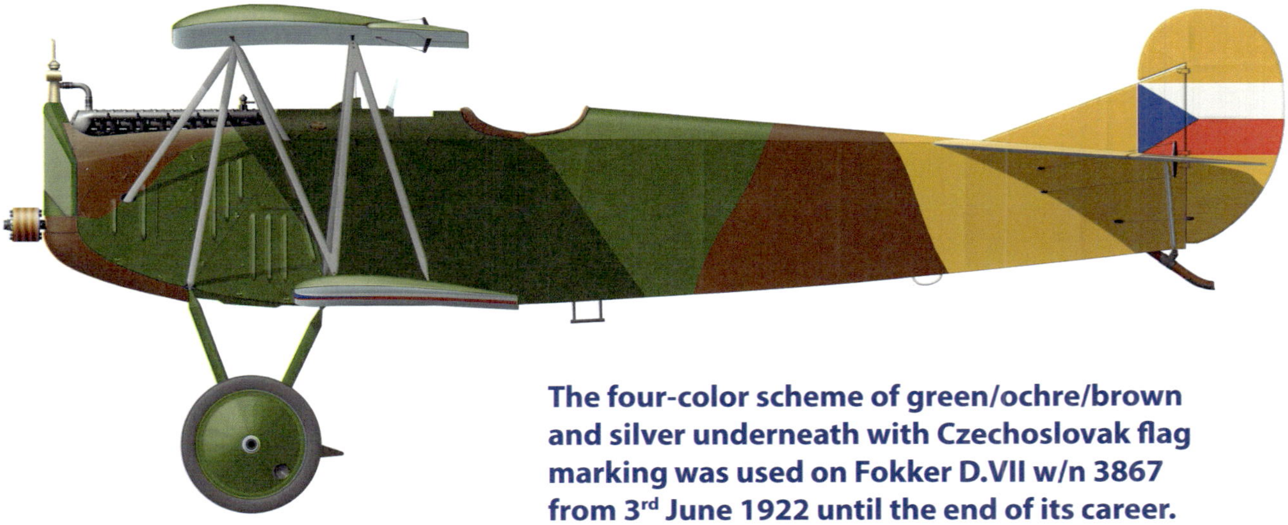

The four-color scheme of green/ochre/brown and silver underneath with Czechoslovak flag marking was used on Fokker D.VII w/n 3867 from 3rd June 1922 until the end of its career. The number 38.67 was repainted on the fuselage; it is visible on a photo dated 29th July 1923 but was later replaced by the marking of the flying school at Cheb (white rectangle)

German Aircraft from the Balkans

Above & Facing Page: Planes and personnel of *Fliegerabteilung* 30 in 1918 on the Salonika Front. (Thorsten Pietsch)

After Romania declared war on Austria-Hungary on 15th August 1916, *Feldmarschall* August von Mackensen – after the successful fighting in Galicia and the occupation of Serbia – was ordered to Bulgaria, where he organized the Danube Army from Bulgarian, Turkish, German, and Austro-Hungarian forces, with which he occupied Dobruja and the Danube delta. Three air units were assigned to the army: FA 28, FA(A) 243, and *Kampfgeschwader* I, and later FA(A) 267 and FA(A) 278 also joined the operations along the Sereth River. By early 1917, the Central Powers' armies had occupied most of Romania, and the Romanians were only managing to hold the areas beyond the Sereth-Danube line with substantial Russian support. At that time, the Bulgarian and German divisions were holding the Macedonian front steady, but in the spring of 1917, the significantly strengthened Entente forces launched a new offensive. In January 1917, 10 squadrons were operating in Macedonia: FA 1, FA 30, FA 34 (ex FFA 66), FA 38 (ex FFA 69), FA(A) 246, FA(A) 271, FA(A) 283, *Jasta* 25, FA Xanthi, and a seaplane station at Kavalla. KG I arrived from Romania in February but was transferred to the Western Front in May. The eight fighters left behind by the unit formed another fighter squadron, Jasta 38. By mid-1917, the following reconnaissance squadrons remained on the southern front of the Balkans: FA 30, FA 34, FA 38, FA 45, FA 66, FA 69, and FA(A) 246.

Brest-Litovsk and Bucharest peace treaties provided some relief for the armies of the central powers fighting in the east. Since reinforcements were badly needed in the west, it is unsurprising that almost all the squadrons from Romania were redirected to the western front. Also, the air units serving in Macedonia could not avoid the downsizing, either. FA 20 (ex FFA 51) was transferred to the Macedonian front to replace the latter ones. During the last year (and especially the last months) of the war, there are records of the operations of six reconnaissance units: FA 20, FA 22, FA 30, FA 34, FA 38, FA(A) 246; two fighter squadrons: Jasta 25, Jasta 38; and the SFA Xanthi. In addition, the supply and resupply of the flying units was the responsibility of Armee-Flug-Park 11. It can be assumed that these squadrons, the air park, and

Above & Below: Bulgarian Roland D.II and German Albatros D.III captured at Krivolak rail station on 21st September 1918. (Boris Ciglić)

Above: Bulgarian POWs preparing a DFW C.V of which was captured by troops of the Danube Division near the village of Slp on 28th September 1918 for transportation. (Boris Ciglić)

their equipment were what left the Balkans with the Mackensen army.

In the aftermath of the Entente offensive in September 1918, the exhausted and ill-equipped Bulgarian armies and the remnants of the German 11th Army could not stop the Entente troops, who had a significant superiority, and the front collapsed within days. Amid the confusion caused by the disorderly return of the soldiers, Bulgaria signed an armistice on 29th September. The retreating German units joined Mackensen's army and set up for defense along the Danube on Romania's southern border. General Kövess wanted to build a defensive line behind the Danube-Sava-Drina rivers. Still, it was evident that neither the Austro-Hungarian nor the German army could stand in the way of the Entente divisions, which were outnumbered and outclassed, and the Monarchy itself was soon to fall to pieces.

The way of most of the aircraft that arrived from the Salonika front ended up in Hungary, but other cases happened, too.

Never Reached Hungary

German air squadrons withdrew with the ground forces. They were already in Serbian territory during October; that is what the German losses suggest at least: five German airmen died in Serbian territory (at Mitrovica, Belgrad, and Niš) between the 6th and 12th of October 1918.

The first two German-built planes (a Bulgarian Roland D.II and a German Albatros D.III) were captured by the Serbian army at the railway station of Krivolak in Macedonia on 21st September 1918, immediately after the breakthrough of the Salonika front, and they also seized nearby a substantial German stock of aviation stuff and fuel. Those two planes were carried back to the storage in Thessaloniki, but there is no information on whether they were later used by the Serbians (they were probably given to the French). On 28th September, the Serbian Danube Division reported that its cavalry squadron captured a chariot with a German airplane west of Slp village, but the Germans also suffered losses in the air during this period. On 13th October, a German airplane was brought down near Aleksinac by infantrymen of 1. Infantry Regiment

Above: The Rumpler C.I of the Kingdom of SHS air service, which was captured on 12th May 1919. (Boris Ciglić)

of the Morava Division, killing both crewmembers. Another aircraft, flying allegedly alongside the first one, crash-landed under the fire of Serbian infantrymen near Belotinac village, in the vicinity of Niš; one of the crew was evacuated in this case. For identifying these casualties, the following German losses are known: *Leutnant d.R.* Alfred Reichmuth and *Unteroffizier* Paul Krüger, pilots of Fliegerabteilung 30, and observer *Leutnant d.R.* Wilhelm Lehmann of Fliegerabteilung 22; all of them were killed in combat operations over Serbia on that date. (Although *Ltn.d.R.* Rechmuth is on the list with Jagodina as the place of loss.)

It is also known that when the Serbian army entered Niš, French units seized a train with eight Fokker D.VII fighters on the stretch between Niš and Pirot, and they later transported these planes to Sofia.

In the autumn of 1918, the Serbian army captured many aircraft and practically everything they could find on airfields in Vojvodina, southern Hungary, and around Temesvár (today Timişoara, Romania). Most of these were Austro-Hungarian aircraft, but there were also some German planes from the Salonika front; their exact number is unknown. In the 1920s, at least three Rumpler C types (two C.IV and a C.VII) were flown operationally by the the Kingdom of SHS air force. They originally belonged possibly to FA 30 or FA(A) 246 and were captured probably at Temesvár. The last mentioned unit had two casualties at Temesvár on 13th November 1918, pilot *Leutnant* Lehmann and observer *Leutnant d.R.* Max Mayer, while a further airman of an unidentified unit died the next day.

Another aircrew lost their lives at Versec on 8th November 1918, but this accident most likely happened a day earlier. The plane of pilot *Vizefeldwebel* Wilhelm Hampe and observer *Leutnant d.R.* August Asemissen of FA 30 crashed near the rail station, killing both airmen. A contemporary Hungarian newspaper article adds more details as it mentions that four German planes took off to leave the town, and one of them crashed during the take-off. Both of the crew were buried in the local cemetery.

One German aircraft that later served in the Kingdom of SHS had come to Serbian territory by another means: a Rumpler C.I was captured in May 1919, and although there are two different versions of the date and circumstances, it happened on 12th May 1919. The Serbian command at Sellye reported an aircraft coming from Pécs passed over Szigetvár at 7.10 p.m., which was fired by a machine gun (from the monitor "*Drina*"/ex S.M.S. "*Temes*", according to some sources), hitting the fuel tank, and forcing the plane to land. The two crewmembers were taken prisoners, while the aircraft was only slightly damaged and appeared to be ready for use soon.

Above: Czechoslovak legionnaires surrounding a landed German LVG C.VI, probably in the late autumn/winter of 1918. (László Winkler)

Passed Hungary

On 9th November 1918, three German airmen lost their lives at Érsekújvár, in what was then Czechoslovak territory, but there is no information on whether they arrived there by plane. Almost a month later, on the 4th December 1918, Četnický *strážmistr* (Gendarme Sergeant) Frantisek Vitula of the Hodonín detachment reported on the following to the higher command:

"On 3rd December 1918 at 4.30 p.m., a biplane was observed circling several times at a height of about five hundred meters above Velka and in the vicinity of Velka. The biplane came from the direction of Hungary. After a longer period, the biplane in question landed in a field near Velká. The undersigned went immediately to the landing site with the probationary gendarme Rudolf Tomanek and four gendarmes as assistance. Corporal Rechtenbach got out of the aircraft as a driver, and Leutnant Valter Schmidt as an observer, both of the aviation department No. 22 in Graudenz, Germany. Both of the airmen mentioned above reported that they had taken off on the same day in S[z] olnok, Hungary, and thought to fly to Bohumín, asking for directions to Brno, were interned and guarded at the local gendarme station. There was one machine gun with 349 rounds of ammunition and 8 illuminating rockets on the aircraft. The aircraft was guarded on the spot after the two aviators were handed over to the headquarters of the Slovak Brigade in Hodonín on 4th December 1918. The aircraft, which was dismantled on 5th December 1918 with the help of a military detachment, was transported to the railway station in Strážnice."

The aircraft LVG C.VI 3981/18 belonged to Fliegerabteilung 22, and its crew consisted of pilot *Unteroffizier* Adolf Rechtenbach and observer *Leutnant* Walter Schmidt. Their mission was to fly the aircraft to Fliegerersatzabteilung 8 in Graudenz. For this flight, they presented a valid document issued by the commander of FA 22 to fly the Szolnok–Gleiwitz–Brieg–Breslau–Graudenz route. The aircraft was impounded, and both airmen received documents for travel to Germany and were released. The plane served later in the Czechoslovak Air Force, mainly at the flight schools in Prague and Cheb, and in 1924 it was disbanded.

Only one German aircraft was operational and flown at a Czechoslovakian frontline air unit: LVG C.VI 4896/18. The designation was mistakenly recorded in Czechoslovakian documents as 4869, and the plane consequently carried 48.69, following the former

Above: LVG C.VI 48.69 of the Czechoslovak *1. polní letecká setnina* (1st field squadron). The original number of the plane (4896) is visible on the tail surfaces. (Petr Lukeš)

Above: Crashed LVG C.VI 4896/18 on the improvised airfield near Losonc (today: Lučenec, Slovakia) on 25th May 1919. The marking of the **"**N" was supposed to indicate belonging to the "Novak" squadron after the commander *setník* Stanislav Novák. (Petr Lukeš)

Above: LVG C.VI 7715/18 landed at Cheb on 22nd June 1919 after their flyers lost their way. The aircraft returned to Germany ten months later. (Petr Lukeš)

Austro-Hungarian designation system. It has been a mystery how this plane got to the country; what is known as the German two-seater was assigned to *1. polní letecká setnina* at Košice on 27th February 1919. Available photos, however, proved that this aircraft was assembled and used at Hungarian unit *8. repülőosztály* in February 1919, which was stationed at Rákos airfield. The squadron's only lost plane was captured on 20th February 1919 by the Czechoslovakians; it is probable that these were the same aircraft. Pilot Kálmán Suták and observer Ottó Polnisch were carrying propaganda material across the demarcation line that day. Still, the Hungarian plane was fired upon over Rimaszombat (today: Rimavská Sobota, Slovakia), and the pilot had to make an emergency landing. The Czechoslovakians captured the Hungarian airmen, and although they did not find any weapons when they searched the plane, only leaflets and copies of the newspaper Népszava, they threatened to execute them. The Hungarian military leadership protested the incident to the Armistice Committee.

At the Czechoslovak squadron *Praporčík* Rudolf Forst carried out the first recorded flight at the control of 48.69 with *nadporučík* Miloslav Doležel on 17th May 1919. During its short service on the airfield of Pezinok near Bratislava and the provisional one at Lučenec, Forst became the privileged pilot of the LVG. It is necessary to remember that the LVG biplane was equipped with a control stick, different from the traditional steering wheel control of the Austro-Hungarian planes, which were flown by the former k.u.k. pilots like Forst and the others.

LVG C.VI 48.69 became out of operation on 25th May when it suffered slight damage on the way back from a reconnaissance flight. The accident occurred at the emergency airfield near Lučenec, and besides the landing gear, the propeller was also damaged. The crew, however, remained unharmed. The airplane could be repaired at the squadron, but they had to wait for the spare parts to arrive from Prague.

On 7th June 1919, the LVG was ready again for a sortie after the landing gear was repaired and with a new propeller. *Četař* Bohuslav Hrazdil was to make a flight, his first one in a plane with a control stick instead of a steering wheel. Hrazdil gave the Benz engine full throttle, and the plane was taking off. At the same time,

Above & Below: German Fokker D.VII 5146/18 with Károly Kaszala at Mátyásföld. Note the marking of the skull and crossbones. (László Winkler & Petr Lukeš)

Above: Captured German DFW C.V at the MÁG factory. (Fortepan)

according to witnesses, the aircraft began to bank to the left at a height of about one meter. The bank became steeper and steeper until the lower left wing touched the airfield's surface. It must have been apparent to the pilot that he could no longer keep the plane under control, and he promptly shut down the engine. The LVG turned 180° around the wing, losing its freshly repaired undercarriage in a flash so that the lower wings absorbed all the remaining energy, in which most of the ribs broke. The petrol tank and propeller also took a hit. The crew, however, miraculously remained unharmed.

Two other LVG C.VIs, 3978/18 and 3984/18, got into Czekoslovakian hands. Their origin is unknown, but they supposedly belonged to a squadron of the Mackensen's army, passing through Czechoslovakia on its way from the Balkan peninsula home to Germany.

Sorties Flown with LVG C.VI 48.69 at *1. polní letecká setnina*

Date	Pilot	Observer	Flying time	Task
19.5.	prap. Rudolf Forst	npor. Miloslav Doležel	2h 10'	recon.
20.5.	prap. Rudolf Forst	npor. Miloslav Doležel	3h 10'	recon.
21.5.	prap. Rudolf Forst	npor. Miloslav Doležel	1h 8'	recon.
22.5.	prap. Rudolf Forst	npor. Miloslav Doležel	2h	recon.
24.5.	prap. Rudolf Forst	npor. František Weinert	2h 15'	recon.
25.5.	prap. Rudolf Forst	npor. Miloslav Doležel	1h 50'	recon.

Above & Below: Special personal markings on the fuselage of a seized LVG C.VI on the airfield of Rákos. (László Winkler)

Above & Following Two Pages: Assembling of LVG C.VIs on the Rákos airfield in February 1919. The serial number 4896 is visible on the wing, which assumes that this plane got into Czechoslovakian hands in the spring of 1919. (László Winkler & Christian Reiterer via Marcus Zelezny)

The First and Last Mention of Czechoslovak LVG C.VIs in Documents

3978/18	first flights at the squadron of Prague on 12th March 1919	written-off on 22nd June 1922 at HLD at Olomouc after a crash in Cheb on 11th May 1922
3981/18	belonged to *Fliegerabteilung* 22, seized after force-landing on 3rd December 1918	31st October 1924 *Vojenské letecké učiliště* (VLU; Military Training Center) at Cheb (permission of Ministry of Defence to write-off)
3984/18	19th February 1920 (reportedly purchased in Germany before this date)	in one and two-seater courses at VLU at Cheb on 21st November 1923
4896/18 (48.69)	delivered to *1. polní letecká setnina* on 27th Feb 1919	burned in a fire at HLD in Prague on 5th November 1921
7715/18	belonged to *Artilleriefliegerstaffel* 116, seized at Cheb after an emergency landing on 22nd June 1919	returned to Bavaria on 12th April 1920 for the promise to stop overflight of the Czechoslovak border

Note: The list does not include two LVG C.VI aircraft or their civil P1 versions, possibly belonging to the Ukrainian Government. Both planes were interned in Czechoslovakia, crashed with Czech pilots, and were later purchased by the Zivnobanka banking house. Their numbers have yet to be discovered; they were often known as the LVG 1 and LVG 2 at Cheb. Both were written off in August 1924.

Captured German LVG C.VI of one of the reconnaissance squadrons belonged to the Mackensen army with unique personal markings on the fuselage

LVG C.VI 4869/18 captured in Hungary in the winter of 1918/1919 and assembled at Rákos airfield in February 1919. The Czechoslovakians recaptured the aircraft and it served at *1. polní letecká setnina* in May 1919

LVG C.VI 7618/18 in Romanian colors after seizing in Hungary in 1919.

Remained in Hungary

As a result of the armistice of 11th November 1918, which ended not only the fighting between the Entente and Germany but the whole of World War I, many of the German airmen headed for home from the Balkan. Already the next day, a German Fokker of Jasta 38 landed at Örkény (50 km south of Budapest) in Hungary, and the landing of another four German aircraft was also reported near Örkény on the same day. On 13th November, the Aviation Department of the Hungarian War Ministry received a report that two cars and two aircraft had been seized from German officers in transit. Two days later, on the 15th, the commander of the National Guard in Bicske (about 35 km west of Budapest) reported the force-landing of a German two-seater due to lack of fuel.

In the Armistice of Belgrade signed on 13th November 1918, the commanding officer of the Allied Army of the Orient, General Louis Franchet d'Esperey, required, among other things, the withdrawal of all German troops from the Balkans within 15 days and the disarmament and internment of German soldiers crossing the country. That was virtually impossible as the Hungarian government did not have sufficient armed forces to compel the Germans to surrender their weapons; it was a challenge even for the returning Hungarian soldiers. Following the repeated French demands, Mackensen finally agreed to disarm and intern his troops in Hungary on 2nd December. The weapons were to be handed over to the railway station commands, after which the consignments were to be forwarded on the instructions of the Hungarian central transport command, the details of which were still to be decided on 2nd December. According to these instructions (quoting only the relevant passages): *"The transports at Szolnok, or to the east or south of it, are to be disarmed at Szolnok, the others at the station nearest to the place of detention. The materials contained in the transports are to be detained at the nearest suitable station to the place of detention and handled by the provisions already issued (...) The removal of German troops in the area of the Arad, Szeged, and Kolozsvár line headquarters remains pending."*

The total number of German troops that passed through Hungary was estimated at 170.000. The first to enter the country were units of the German 11th Army stationed in the Vojvodina region. Mackensen's troops began to evacuate Romania on 10th November 1918, and the first units reached the Hungarian border on 19th November. Most German troops were transported by

rail, with the last German transport crossing the border and leaving the country on 3rd January 1919. It is worth noting that the German troops had, in the meantime, found their way back, and they, in many cases, scattered with the knowledge of their commanders. The retreat of the Mackensen army did not take place in complete military order, and the soldiers often tried to return home by selling their surplus military equipment.

In reality, the disarmament occured as follows (according to a contemporary newspaper article): *"At first, on the instructions of the Hungarian government, the station commanders called on the marchers to lay down their weapons in a friendly manner, but no violence was used. The German troops did not comply with the order until Saturday of last week* [2nd December 1918], *when the so-called disarming committee of the military council of the Mackensen army appeared in Szolnok, and from then, it disarmed all incoming transports. The Germans thus disarmed themselves without the slightest incident or disturbance. The disarmament committee handed over the weapons and munitions to the station command at Szolnok, which took custody of them. This continued until Wednesday morning. At that time, Mackensen entered into negotiations with the Hungarian government concerning the details of the disarmament and possible internment, and then, yesterday* [4th December 1918], *at the call of the Hungarian government, there was a pause in the disarmament at Szolnok. The Hungarian government sent a three-person committee to Szolnok to assist in the disarmament under the new detailed agreement."*

Mackensen, seeing that neither the Hungarian government nor the Entente was able to prevent his army's march, made every effort to avoid disarmament and internment. On 5th December, he ordered the German troops that "disarmament and planned internment were suspended (...) Valuable material (...) must be removed".

The Hungarian air disarmament committees were formed on 7th December 1918, and were soon up and running. Although the Entente demanded the surrender of all weapons, including aircraft, the Hungarian side failed to do so: the German war material delivered to the Entente until the proclamation of the Hungarian Soviet Republic contained practically no air material (except for two hangars). Several reminiscences stated, however, that German aircraft were standing in closed wagons

Above: LVG C.VI with an observer machine-gun in the Hungarian Red Flying Troops' service. (Fortepan)

on the rails when the authorities seized them, and a significant part (or maybe all) of them were transported from Jászapáti, according to the records at least.

The captured German planes were continuously carried to the Material Depot at Cinkota (now part of Budapest). Based on the Material Depot's inventories, there were ten aircraft of German origin on 5th January 1919 and 20 aircraft on 1st February 1919. Later, the number of German aircraft increased significantly, with 63 aircraft registered on 1st April and 53 on 31st May.

Only minimal information is available about the airplanes of German squadrons that served on the Salonika front, especially in 1918 and even more so in the last months of the war. Gerhard Fieseler, ace of Jasta 25, wrote for instance, that in early July 1918, two Fokker D.VIIs were sent to the Macedonian theatre. He got one, and the other went to *Leutnant* Fritz Thiede, CO of Jasta 38.

The German planes that fell into Hungarian hands were Albatros D.Va and Fokker D.VII fighters, DFW C.V, Halberstadt C.V, LVG C.VI, Pfalz C.I and Rumpler C.VI/C.VII reconnaissance planes, most in a more or less damaged condition. Indeed, some aircraft did not arrive at Cinkota or are not listed in the records. As the Hungarian Flying Troops needed a constant supply of aircraft, especially during the fighting of the Hungarian Soviet Republic, the command decided to deploy German reconnaissance and fighter aircraft in addition to ones produced by the Hungarian aircraft companies. The primary German "collection" types of about 70 aircraft were the Fokker D.VII fighter and the LVG C.VI reconnaissance aircraft. They used fewer of the former, probably because MÁG also produced the type (series 93), but the final assembly of six brand new German Fokker D.VIIs was also carried out there. The newly built fighters had an obvious advantage over the stripped-down German aircraft. The situation was different with the LVG reconnaissance aircraft, as they were worthy counterparts of the domestically produced aircraft in terms of performance and reliability. Approximately 30 of this type were in Hungarian hands, ten of which were undoubtedly flown. In May 1919, more aircraft were transferred from Cinkota to Mátyásföld (MÁG) and Albertfalva (UFAG; MARE in

German Aircraft in Hungary				
5.1.1919		1.2.1919		
Serial No.	Engine No.	Serial No.	Engine No.	Condition
DFW C.V				
		2199/17	30839	not operational
Fokker D.VII				
322/18	- (32591)	322/18	-	repairable
4053/18	31118	4053/18	31118	deflected tail-end
4067/18	33059	4067/18	33059	damaged, repairable
7606/18	3458(?)	5061/18	42814	slightly damaged
		5141/18	33045	good, slightly damaged
		5287/18	-	fabric ripped off
		5319/18	-	fabric ripped off
		5323/18	-	fabric ripped off
		5328/18	42978	fabric ripped off
		5329/18	-	fabric ripped off
Halberstadt C.I (DFW-Built)				
		C4141/18	34898	no operational
		C4145/18	31682	no operational
Halberstadt C.V				
		3458/18	34900	slightly damaged
		3459/18	34990	damaged, repairable
		3465/18	35015	slightly damaged
LVG C.VI				
3959/18	34484	3959/18	33484	repairable
3979/18	32392	3979/18	32392	repairable
8921/18	35377			
Rumpler C.IV/C.VII				
3347/18	40066	3347/18	52383	good
4056/18	27571	4056/18	27571	badly damaged
Unknown Type				
C/6461	31877			

Hungarian) for repairs.

First *8. repülőosztály*, based at Rákos, was equipped with LVG biplanes, two of which were assigned to the unit. One of them was brought from the Material Depot at Cinkota to the UFAG at Albertfalva, where it was assembled and then flown to Rákos (probably one with serial number 8921/18). At the same time, the other one was delivered directly to the unit commanded by Sándor Hartzer. One of the missions flown by the German LVG C.VIs was recalled by Gyula Troján, former pilot of Flik 58 and Flik 41/J:

"This is how I flew in mid-March 1919 in a German plane of type LVG equipped with a 300 hp engine over the then surrounded Salgótarján and Kassa. We were carrying mail and leaflets, which my observer officer dropped. We were greeted with heavy gunfire at both places, but flying low, we accomplished our task, and after a strong five-hour flight, we returned safely to Rákos. *It was impossible to fly a*

Above: An LVG C.VI of *6. repülőszázad* on the airfield at Onga before taking-off for a sortie in the summer 1919. (László Winkler)

Hungarian plane for that long because it did not have such a large fuel tank."

Later, *2. repülőosztály/repülőszázad* at Albertfalva also received some of the excellent German reconnaissance aircraft, but LVG C.VIs were used in the most significant number by the *6. repülőszázad* in the summer of 1919. The unit under the command of Béla Schwéger was stationed at Onga near Miskolc. A field east of the village was used as an airfield; it was the pasture of the village, and the flat, grassy area was ideally suited to the purpose of the temporary airfield. No structures were erected on the site, aircraft were probably stored in hangar tents, and most of the crew were accommodated in the nearby castle. The squadron's main task was reconnaissance in an easterly and northerly direction, but enemy troop movements were also to be disrupted by bombing.

The LVG C.VIs proved reliable, and the pilots liked the type, but regular reconnaissance missions wore out the planes well. On 18th June, for example, after the engine of the only LVG still in service "ran hot", there was no operational aircraft for about a day and a half. The next day's report highlighted the shortages: "Only two barrels of petrol and one semi-airworthy aircraft in the *6. repülőszázad*. Petrol and the 1 LVG and 2 Berg planes promised yesterday would be needed."

The unit's aircraft continued to suffer regular damage afterward. On the afternoon of 21st June, Mihály Kimmel and Alfréd Schuff, flying aircraft 7711/18, were forced to abort their reconnaissance mission due to dense fog and engine failure, and the plane suffered minor damage on landing. On the morning of 1st July, Béla Bodó and Antal Kristolofil departed to reconnaissance the line of the Tisza River and the area around Debrecen, but during the flight, an anti-aircraft battery from the south of Hajdúnánás opened fire on them. LVG C.VI 8952/18 was hit, most probably the propeller, which caused it to fly away, and the pilot made a force landing near Csanálos-Sóstó. On the morning of 5th July, Miklós Lipcsey and László Diószeghy crashed 8951/18 on landing after their mission. On 13th July, Lajos Breier and Kristolofil brought 7711/18 home with two hits on the left wing surface after heavy artillery fire over

Captured German Planes in the Material Depot at Cinkota

Serial No.	Engine No.	Arrived at Cinkota		Condition	Notes
		Date	From		
Albatros D.Va					
6989/17	30918	2–3.1919	Jászapáti	badly damaged	OAW-built
6993/17	31111	2–3.1919	Jászapáti	badly damaged	OAW-built
6994/17	32040	2–3.1919	Jászapáti	badly damaged	OAW-built
6995/17	31008	2–3.1919	Jászapáti	badly damaged	OAW-built
6996/17	31565	2–3.1919	Jászapáti	badly damaged	OAW-built
6997/17	31451	2–3.1919	Jászapáti	badly damaged	OAW-built
6998/17	26785	2–3.1919	Jászapáti	slightly damaged	OAW-built
7507/17	32811	2–3.1919	Jászapáti	badly damaged	
DFW C.V					
2199/17	30939	1.1919		badly damaged	LVG-built Engine disassembled: 18.3.1919
3834/17	30078	2–3.1919	Jászapáti	badly damaged	Aviatik-built
274/18	30515	2–3.1919	Jászapáti	badly damaged	Aviatik-built
Fokker D.VII					
322/18	-	12.1918		slightly damaged	dual-control
598/18	32752	2–3.1919	Jászapáti	badly damaged	Albatros-built
3247/18*	35682	---			
4053/18	34118	12.1918	33484	badly damaged	OAW-built
4067/18	33059	12.1918	32392	slightly damaged	Engine disassembled: 20.5.1919
4072/18*	35416	---			OAW-built
5061/18	42840	1.1919	52383	good	
5141/18	33045	1.1919	27571	badly damaged	
5146/18*	?	---			
5207/18	42306	2–3.1919	Jászapáti	slightly damaged	Albatros-built
5209/18	42452	2.1919	Jászapáti	slightly damaged	Albatros-built Engine disassembled: 3.3.1919
5287/18	42814	1.1919		slightly damaged	Albatros-built
5319/18	43136	1.1919		slightly damaged	Albatros-built
5323/18	42806	1.1919		slightly damaged	Albatros-built
5328/18	43978	1.1919		slightly damaged	Albatros-built
5329/18	43038	1.1919		slightly damaged	Albatros-built
7606/18	3458?	12.1918		?	

Captured German Planes in the Material Depot at Cinkota (continued)					
Serial No.	Engine No.	Arrived at Cinkota		Condition	Notes
		Date	From		
Halberstadt C.V					
3458/18	34900	1.1919		good	
3459/18	34990	1.1919		slightly damaged	Engine disassembled: 21.5.1919
3465/18	35015	1.1919		good	
4141/18	34898	1.1919		for rejection	DFW-built Rejected: 4.1919
4145/18	31682	1.1919		for rejection	DFW-built Rejected: 4.1919
LVG C.VI					
3346/17**	30386	2–3.1919	Jászapáti	badly damaged	*LVG C.V*(?)
3959/18	34484	12.1918		slightly damaged	Shipped to Mátyásföld (repair): 10.5.1919
3979/18	32392	12.1918		slightly damaged	
3997/18	32754	2–3.1919		badly damaged	
6372/18*	24726	---			
6721/18*		---			
6751/18*		---			
6961	-	2–3.1919		for rejection	Rejected: 4.1919
7164/18	32015	2–3.1919		badly damaged	
7486	33279	2–3.1919	Jászapáti	slightly/badly damaged	
7491/17**	32513	2–3.1919	Jászapáti	slightly/badly damaged	
7494	37111	2–3.1919	Jászapáti	slightly/badly damaged	
7601/18	34704	2–3.1919		slightly damaged	Shipped to Mátyásföld (repair): 10.5.1919
7605/18	35033	2–3.1919		slightly damaged	
7608/18	35032	2–3.1919		slightly damaged	
7611/18	35069	2–3.1919		good	Shipped to Mátyásföld (repair): 10.5.1919
7618/18	33185	2–3.1919		slightly damaged	
7620/18	34984	2–3.1919		slightly damaged	
7636/18	35073	2–3.1919		slightly damaged	Shipped to Mátyásföld (repair): 10.5.1919
7638/18	35042	2–3.1919		slightly damaged	Shipped to Mátyásföld (repair): 10.5.1919
7646/18	35077	2–3.1919		slightly damaged	
7699/18	30698	3.1919	Nyíregyháza	badly damaged	
7706/18	35147	26.4.1919		slightly damaged	
7711/18	22444	2–3.1919		good	
7713/18	22314	2–3.1919		slightly damaged	

Captured German Planes in the Material Depot at Cinkota (continued)					
Serial No.	Engine No.	Arrived at Cinkota		Condition	Notes
		Date	From		
7791/18*	?				
8921/18	35377	12.1918		slightly damaged	
8951/18	35481	2–3.1919		slightly damaged	
8952/18	35355	2–3.1919		slightly damaged	
8966/18	35380	2–3.1919		slightly damaged	
8974/18	35352	2–3.1919		good	Shipped to Albertfalva (repair): 10.5.1919
8987/18	35432	2–3.1919		good	
Pfalz C.I					
7545/16**	-	2–3.1919		slightly damaged	
7555/18	36588	2–3.1919		slightly damaged	
7556/18	38185	2–3.1919		good	
Rumpler C.IV/C.VII					
3346/18	-			badly damaged	
3347/18	40066	12.1918		slightly damaged	Engine disassembled: 19.5.1919
4035/18	-	2–3.1919		for rejection	Rejected: 4.1919
4056/18	33059	12.1918		badly damaged	
Undentified Types					
2300/18	-	1.4.1919		for rejection	*DFW C.V(?)* Rejected: 4.1919
2800/18	-	2–3.1919		for rejection	*Halberstadt Cl.IIa(?)* Rejected: 4.1919
6461	31877	2–3.1919		?	

Note: some serial numbers are incomplete in the original lists, missing the year of production.
* serial numbers are not included in the inventories of the Material Depot, based on other sources.
** serial numbers with probably wrong year of production.

Known Sorties of *8. repülőosztály* Flown by German LVG C.VIs					
Date	Pilot	Observer	Aircraft	Route	Task
10.2.	István Deutsch	Miklós Rohla	6721/18	Rákos–Vác–Pásztó–Miskolc–Rákos S: 13.00 L: 14.50	prop.
10.3.	Gyula Troján	Miklós Rohla	6751/18	Aszód–Salgótarján–Rimaszombat–Bátka–Tornalja–Pelsőcz–Sziliceszádalmás–Tomagörgő–Torna–Somodifürdő–Szepsi–Csécs–Szurdok–Hernád–Miskolc–Maklár–Gyöngyös–Hatvan–Rákos S: 10.30 L: 14.30	prop.
12.3.	László Szabó	Sándor Neogrády	LVG	Rákos–Feled–Rimaszombat–Sajó–Gömör–Verpelit–Rákos (Murányig Bánréve–Pelsőcöt)	prop.

Known Sorties of *2. repülőosztály/2. repülőszázad* Flown by German LVG C.VIs					
Date	**Pilot**	**Observer**	**Aircraft**	**Route**	**Task**
20.4.	Géza Csenkey	Miklós Rohla	LVG	Albertfalva–Szolnok–Debrecen	carrying leaflets
7.6.	Károly Debreczeni	Dezső Smerd	LVG	Albertfalva–Szolnok–Fülöpszállás–Albertfalva S: 7.00 L: 9.35	recon.
10.6.	István Szöllösy	Dezső Smerd	LVG	Albertfalva–Balassagyarmat–Losonc–Albertfalva	recon.

Known Sorties of *6. repülőszázad* Flown by German LVG C.VIs					
Date	**Pilot**	**Observer**	**Aircraft**	**Route**	**Task**
8.6.	Géza Tauth	Pál Örvös	8987/18	Onga–Tiszadob–Tiszaluc–Bűdszentmihály–Tiszalök–Nyíregyháza–Rakamaz–Onga S: 17.05 L: 18.45	recon.
10.6.	Andor Bauer	Pál Örvös	7711/18	Onga–Kassa–Kassabéla–Margitfalva–Korompa–Hernád völgye–Onga S: 6.45 L: 8.25	recon.
11.6.	Lajos Breier	Antal Kristolofil		Onga–Nyíregyháza–Csap–Onga	recon.
12.6.	Miklós Lipcsey	Antal Kristolofil	8987/18	Onga–Sátoraljaújhely–Királyhelmet–Csap–Ungvár–Nagymihály–Köketerebes–Sátoraljaújhely–Királyhelmet–Nagymihály–Onga S: 14.00 L: 16.30	recon.
14.6.	Waldemár Renner	György Zachariás	LVG	Onga–Szerencs–Sátoraljaújhely–Ungvár–Onga S: 17.50 L: 18.10	recon.
15.6.	Waldemár Renner	György Zachariąs	8987/18	Onga–Nagymihály–Ungvár–Csap–Királyhelmec–Onga S: 7.05 L: 9.25	recon.
16.6.	Waldemár Renner	Antal Kristolofil	8987/18	Onga–Tiszadob–Budszentmihály–Hajdúnánás–Hajdúdorog–Hajdúböszörmény–Debrecen–Nyíregyháza–Rakamaz–Onga S: 18.00 L: 20.30	recon.
17.6.	Géza Tauth	István Reviczky	7711/18	Onga–Tiszalúc–Szederkény–(Csap–Debrecen)–Kisfaludy puszta–Tiszapolgár–Hajdnánás–Kisszentmihály–Rakamaz–Tiszalök–Tiszadada–Tiszadob–Onga S: 17.15	recon.
18.6.	Mihály Kimmel	Alfréd Schuff	8952/18	Onga–Kassa–Onga	recon.
	Andor Bauer	György Zachariás	7711/18	Onga–Tokaj–Balsa–Karád–Szentmáriamajor–Rakamaz–Onga S: 19.10 L: 20.15	recon.
21.6.	Mihály Kimmel	Alfréd Schuff	7711/18	Onga–Sátoraljaújhely–Tőketerebes–Nagymihály–Onga S: 14.30 L: 17.30	recon.
22.6.	Lajos Breier	István Reviczky		Onga–Miskolc–Nagyrőce–Csetnek–Rozsnyó–Szomolnok–Szepesremete–Onga S: 6.55 L. 9.05	recon.
23.6.	Ferenc Maurer	Pál Örvös	8951/18	Onga–Sátoraljaújhely–Tőketerebes–Gálszécs–Varannó–Nagymihály–Onga S: 16.00 L: 18.55	recon.

Known Sorties of *6. repülőszázad* Flown by German LVG C.VIs (continued)

Date	Pilot	Observer	Aircraft	Route	Task
1.7.	Béla Bodó	Antal Kristolofil	8952/18	Onga–Taktakenéz–Tiszadada–Tiszapolgár–Csege–Balmazújváros–Debrecen–Hajdúböszörmény–Hajdúdorog–Hajdúnánás–Bűdszentmihály–Tiszalök–Onga S: 10.00 L: 12.20	recon.
5.7.	Miklós Lipcsey	László Diószeghy	8951/18	Onga–Tiszapolgár–Debrecen–Nyíregyháza–Rakamaz–Onga S: 8.10 L: 10.10	recon.
6.7.	Béla Bodó	György Zachariás	7711/18	Onga–Szabolcs–Tiszafüred–Rakamaz–Csőszhalom–Tiszaeszlár–Debrecen–Nyíregyháza–Balmazújváros–Orosháza–Onga S: 15.30 L: 17.15	recon.
10.7.	Andor Bauer	Alfréd Schuff	7711/18	Onga–Szerencs–(Losonc)–Tokaj–Vencsellő–Tiszaeszlár–Onga S: 7.00 L: 8.32	recon.
11.7.	Mihály Kimmel	István Reviczky	7711/18	Onga–Tiszaeszlár–Vencsellő–(Rakamaz, Királytelek)–Onga S: 7.05 L: 8.20	recon.
12.7.	Ferenc Maurer	Béla Beregszászi	7711/18	Onga–Tokaj–Tímár–Szabolcsbalcsa–Királytelek–Nagyfalu–Rakamaz–Onga S: 6.35 L: 7.25	recon.
13.7.	Lajos Breier	Antal Kristolofil	7711/18	Onga–Tokaj–Tímár–Rakamaz–Nagyfalu–Szederkény–Tiszaeszlár–Onga S: 7.25 L: 8.45vm	recon.
15.7.	Béla Bodó	László Diószeghy	7711/18	Onga–Tokaj–Rakamaz–Tiszaeszlár–Onga S: 5.50 L: 7.20	recon.
	Andor Bauer	Antal Kristolofil	7711/18	Onga–Tokaj–Rakamaz–Balsa–Tímár–Tokaj–Onga S: 16.20 L: 17.35	recon.
18.7.	Géza Tauth	Pál Örvös	7711/18	Onga–Rakamaz–Nyíregyháza–Debrecen–Onga S: 7.30 L: 9.30	recon.
19.7.	Waldemár Renner	Béla Beregszászi	8966/18	Onga–Rakamaz–Tímár–Szabolcs–Balsa–Nagyfalu–Onga	recon.
20.7.	Lajos Breier	Pál Örvös	7711/18	Onga–Tiszapolgár–Hajdúnánás–Nyíregyháza–Balsa–Rakamaz–Onga S: 18.55 L: 20.25	recon.
21.7.	Béla Bodó	László Diószeghy	8966/18	Onga–Tiszalök–Tiszadada–Hajdúnánás–Nyíregyháza–Tokaj–Onga S: 7.40 L: 9.10	recon.
	Dezső Jäger	László Antal	7711/18	Onga–Tarcal–Nagyfalu–Rakamaz–Tímár–Balsa–Vencsellő–Onga S: 10.50 L: 12.50	recon.
	Géza Tauth	György Zachariás	7711/18	Onga–Balsa–Vencsellő–Gáva–Tiszanagyfalu–Gémeskút–Tiszaeszlár–Tiszalök–Onga S: 18.10 L: 19.10	recon.
23.7.	Andor Bauer	Pál Örvös	7711/18	Onga–Tokaj–Rakamaz–Nagyfalu–Nyíregyháza–Kisvárda–Csap–Onga S: 11.00 L: 13.30	recon.
24.7.	Mihály Kimmel	István Reviczky	7646/18	Onga–Tarcal–Újszántóhalom–Királytelek–Onga S: 17.05 L: 19.10	recon.
	Waldemár Renner	László Diószeghy	7711/18	Onga–Csap–Gara–Tiszalök–Onga S: 18.30 L: 19.05	recon.

Known Sorties of *6. repülőszázad* Flown by German LVG C.VIs (continued)					
Date	**Pilot**	**Observer**	**Aircraft**	**Route**	**Task**
24.7.	Mihály Kimmel	István Reviczky	7646/18	Onga–Tarcal–Újszántóhalom–Királytelek–Onga S: 17.05 L: 19.10	recon.
	Waldemár Renner	László Diószeghy	7711/18	Onga–Csap–Gara–Tiszalök–Onga S: 18.30 L: 19.05	recon.
25.7.	Lajos Breier	György Zachariás	7711/18	Onga–Bodrogkeresztúr–Karád–Káva–Csőszhalom–Rózsástanya–Onga S: 7.30 L: 8.55	recon.
	Béla Bodó	Pál Örvös	7711/18	Onga–Nyíregyháza–Csap–Onga S: 17.00 L: 17.50	recon.
28.7.	Géza Tauth	Antal Kristolofil	7711/18	Onga–Dorogma–Tisza–Tokaj–Onga S: 10.30 L: 12.10	recon.
	Waldemár Renner	István Reviczky	7648/18	Onga–Mezőcsát–Szeged–Tisza–Tokaj–Onga S: 15.00 L: 16.35	recon.
29.7.	Dezső Jäger	György Zachariás	–	Onga–Takta–Párkány–Kenéz–Tiszaluc–Onga S: 7.25 L: 8.10	recon.
	Andor Bauer	Antal Kristolofil	7946/18	Onga–Tiszadada–Tiszaeszlár–Onga S: 7.40 L: 8.40	recon.
	Géza Tauth	István Reviczky	7711/18	Onga–Tiszadob–Tiszadada–Taktakenéz–Szabolcs–Kenézlő–Balsa–Rakamaz–Onga S: 6.00 L: 7.15	recon.
	Andor Bauer	György Zachariás	7646/18	Onga–Szederkény–Kisfaludipuszta–Tiszafalbomja–Tiszapolgár–Rakamaz–Onga S: 6.40 L: 7.40	recon.
	Dezső Jäger	György Zachariás	7711/18	Onga–Miskolc–Zsolca–Diósgyőr–Onga S: 10.40(11.05) L: 11.50	pursuit enemy plane
	Andor Bauer	Antal Kristolofil	7711/18	Onga–Tiszakies–Tisza–Tokaj–Balsa–Onga S: 15.35 L: 17.15	recon.
30.7.	Dezső Jäger	György Zachariás	7711/18	Onga–Taktaszada–Prügy–Tokaj–Zárold–Mezőzombor–Taktaharkány–Taktakenéz–Tiszalúc–Onga S: 9.00 L: 9.50	recon.
31.7.	Andor Bauer	Antal Kristolofil	7711/18	Onga–(Tiszadada–Tokaj)–Taktakenéz–Prügy–Báj–Csobaj–Tardos–Tiszalök–Rakamaz–Onga S: 7.45 L: 10.00	recon.

Rakamaz. The other side of the balance, however, was several successful sorties. Out of the 48 known sorties flown by the squadron during June and July 1919, at least 41 were flown by LVG C.VI aircraft.

From 17th July onwards, aerial reconnaissance of the squadron had to be limited to accurately identifying of positions of the enemy batteries in preparation for the planned offensive on the Tisza and cooperating with the artillery. Before the each mission started, the company commander had to request reconnaissance instructions from the competent artillery brigade headquarters. Although the unit carried out several reconnaissance sorties almost every day when weather permitted it, and there were occasions when one or two operational aircraft were in the air nearly all day, the higher command was not satisfied. On 28th July, the following order was sent to the squadron: *"After receiving the order to take off, several hours always elapse before, after repeated urging, the aircraft assigned to reconnaissance finally takes off. In the present situation, however, an urgent reconnaissance of a district is usually required, which in this form becomes directly impossible. I, therefore, order that, from tomorrow onwards, the planes capable of taking off shall be prepared in such a way that, no later than five minutes after receiving the order to take off, the plane assigned to reconnaissance can take off."*

On 29th July, the the squadron's pilots made five take-offs, four of which were near the crossing points on the Tisza. In the morning, however, an LVG C.VI 7711/18 and its crew took to the air in pursuit of an intruder: an aircraft reported as a French(?) one heading from Debrecen to Onga, via Miskolc to Izsó was continuing its flight despite heavy machine gun fire. The Hungarian two-seater chased it to Miskolc, but here, pilot Dezső Jäger and observer György Zachariás lost sight of the enemy aircraft in the clouds.

German Aircraft at Hungarian Units

Hungarian Flying Troops (November 1918 – 21st March 1919)

1. repülőosztály (Mátyásföld):

Fokker D.VII 3247/18 (E: 35682)
Fokker D.VII 4072/18 (E: 35416)
Fokker D.VII 5146/18 (E: 35628)

All three were on the Department of Aviation's list dated 3rd February 1919 as aircraft to be allocated to the unit. It is unknown if they were used, and they were not listed in any subsequent registry or account, either.

2. repülőosztály (Albertfalva):

LVG C.VI 6372/18 (E: 24726)

It was on the list of the Department of Aviation dated 3rd February 1919 as aircraft to be allocated to the unit. The plane was flown several times in February 1919.

8. repülőosztály (Rákos):

LVG C.VI 6721/18
LVG C.VI 6751/18

Both served and were flown at the unit in February–March 1919.

Hungarian Red Flying Troops (21st March – 1st August 1919)

2. repülőszázad (ex *2. repülőosztály*, Albertfalva):

LVG C.VI 6372/18 (E: 24726)

At the unit: April–July 1919, it was deployed several times but was under repair for extended periods.

3. repülőszázad (ex *8. repülőosztály*, Rákos):

Halberstadt C.V 3459/18 (E: 34990)

It was sent to the squadron on 21st May 1919, but no other information on its use.

6. repülőszázad (Onga):

LVG C.VI 7646/18 (E: 35077)

At the unit: 24–31 July 1919, flying four known sorties. It was under repair on 27th and 31st July 1919.

LVG C.VI 7711/18 (E: 22444)

At the unit: 8th June – 31st July 1919, flying 25 known sorties. The aircraft's engine "ran hot" on 18th June 1919. The plane was damaged on landing on 21st June 1919. It was under repair between 22nd June and 1st July 1919, and 7–9 July 1919 again. Two anti-aircraft hits on the left wing on 13th July 1919.

LVG C.VI 7713/18 (E: 22314)

At the unit: 29th June – 20th July 1919, flying no known sortie. It was under repair between 29th June and 3rd July 1919, and 5–19 July 1919 again.

LVG C.VI 8951/18 (E: 35481)

At the unit: 23rd June – 11th July 1919, flying two known sorties. The plane crashed on landing on 5th July 1919 and was under repair 5–11 July 1919.

LVG C.VI 8952/18 (E: 35355)

At the unit: 12th June – 27th July 1919, flying two known sorties. It was under repair 12–14 June 1919. The plane's propeller was hit and crash-landed on 1st July 1919. It was under repair 2–7 July 1919 and 9–23 July 1919.

LVG C.VI 8966/18 (E: 35380)

At the unit: 15–23 July 1919, flying two known sorties. It was under repair 15–16 July and on 23rd July 1919.

LVG C.VI 8987/18 (E: 35432)

At the unit: 8–22 June 1919, flying four known sorties. It was under repair 12–13 June and 17–22 June 1919.

7. repülőszázad:

Fokker D.VII 7606/18

It fell into Romanian hands in Debrecen on 23rd April 1919. It was under repair in June 1919 at the Romanian *Escadrila - depozit*

LVG C.VI 7791/18

Captured by Rumanian forces in Debrecen, flown by Escadrila N.7 in June 1919.

Back to the Balkans

After the unsuccessful crossing of the Tisza River, the Eastern Front collapsed, and the Hungarian troops began a full retreat. On 1st August, the Governing Council resigned, ending the period of the Hungarian Soviet Republic, and three days later, Romanian troops marched into Budapest. This was the beginning of a process that, from the Romanian point of view, was a capture or prize, but from the Hungarian point of view, it was a plunder of the country and the aviation industry. This process can be reconstructed with a fair degree

Left: Local ladies posing in front of an LVG C.VI somewhere in Hungary. (László Winkler)

of accuracy based on the situation reports received by the Air Department of the Hungarian War Ministry. According to the data up to 21st November 1919, 302 wagons of flying materials were taken by the Romanians from various places, of which 97 wagons had already left the country via the Tisza bridges between 26th September and 21st November 1919. The War Ministry account shows that Romanians carried away 547 aircraft (204 without engines) and 135 engines altogether. Based on archival sources, Valeriu Avram stated that the Romanian Army captured ca. 150 aircraft of different types (ca. 80 were in good condition) and about 260 aircraft engines in Hungary in 1919. This considerable contradiction is unlikely to be fully resolved, but research is ongoing to clarify the data and paint a more complete picture.

As for the aircraft of German origin, Romanian *Grupul V Aviatie* got the first ones already in the beginning phase of the campaign against Hungary. Occupying Debrecen, they got at least one Fokker D.VII (7606/18) and one LVG C.VI (7791/18); the former appeared at *Escadrila - depozit* on 25th May 1919, then assigned to *Escadrila N.7*, the latter was also flown at *Escadrila N.7* in June. At the end of June 1919, *Escadrila S.2* was equipped with another LVG with serial number 14640 (?).

Above: László Diószeghy of *6. repülőszázad* in the pilot's seat of an LVG C.VI. (Reinhard Zankl)

During the retreat from the Romanian Army, *6. repülőszázad* reached Gödöllő (about 30 km northeast of the outskirts of Budapest) and disembarked here. Still, as the further route was blocked, the entire unit's equipment fell into enemy hands, including four operational and two dismounted aircraft. (On 31st July, the squadron had LVG C.VI 7711/18 in operational condition and 7646/18 under repair, besides three Hansa-Brandenburg C.I series 269.) Identification of further German planes is possible in the case of LVGs, as there are only suitable Hungarian and Romanian sources about them. According to Hungarian reports, three LVG C.VIs were taken from UFAG and nine from MÁG. Valeriu Avram, however, found ten LVG C.VI in Romanian records: 7601/18, 7605/18, 7618/18, 7620/18, 7646/18, 7791/18, 8904/18, 9821/18, 14640 (?) and 15844 (?); three of them were captured earlier as it was already mentioned. From Romanian documents a further LVG is known: 9484/18.

The captured aircraft, especially the LVG C.VIs, entered Romanian service mostly in 1920, and as far as the fate of the planes is concerned, we have information mainly about their accidents. The first happened in 1919 at Debrecen, with *Capitan* Gheorghe Racoveanu in the pilot's seat. 7618/18 had an accident sometime in 1920 in Iaşi piloted by *Sublocotenent* Moşu Simionescu. 7636/18 also suffered damages in Brasov on 17th August 1920 with *Adjutant* Schwartz and *Sublocotenent* Dimitriu.

On 13th September 1920 *Locotenent* Friedrich Rozstocki took off with Nr. 15884 (its original number is unidentified) when the engine, at 50 meters high above the airfield, began to give increasing failures due to the carburetor, forcing it to land with the backward vault, not having time to take it to the front. The propeller, two wheels, and the undercarriage were destroyed, but the engine, the fuselage and the upper wings remained good.

The following day, No. 1920 L (7620/18) crashed at the *Centrul de Instrucţie al Aviaţiei* (Aviation Training Center) in Siret with pilot *Locotenent* Traian Pâclea and pilot student *Locotenent* Alexandru Cernescu. The aircraft was heavily damaged: the wings and the propeller were broken into several pieces, as well as the fuselage, which was broken into three pieces. The engine, like the onboard devices, was destroyed, as were the wheels and the undercarriage. The plane became useless, and even spare parts could not be taken from it as they were all destroyed.

The last known accident happened on 15th August

Above: Rudolf Blass in an LVG C.VI. Note the fixed machine gun within the pilot's reach. (László Winkler)

Above: LVG C.V 9667/17 on the Kiev airfield. Russian pilot Victor Hodorovich arrived in Mátyásföld with this plane on 15th April 1919.

Right: Aircraft wings awaiting demolishing in 1921 after the order of the International Aeronautical Control Commission. Note the wing with the German serial number 4556. (László Winkler)

Right: *Locotenent* Gheorghe Stâlpeanu beside an LVG C.VI. (Dan Antoniu)

1921, when *Locotenent* Petre Davidescu landed with 7646/18, which suffered severe damage to the wheels, the undercarriage, and the lower wing; its propeller was broken, and the engine was displaced.

Above & Below: *Capitain* Racoveanu's accident in Debrecen in 1919. (Dan Antoniu)

Above: LVG C.VI 7618/18 after the landing in Iasi in 1920. (Dan Antoniu)

Above: Crash-landing of *Adjutant* Schwartz and *Sublocotenent* Dimitriu in Brasov on 17th August 1920 with 7636/18. (Dan Antoniu)

Above & Below: *Locotenent* Friedrich Rozstocki's force-landing on 13th September 1920. (Dan Antoniu)

Above, Below, & Following Page: Photo series of 7620/18 after its crash on 14th September 1920. (Dan Antoniu)

Below: Accident of LVG C.VI 7646/18 on 15th August 1921 (Dan Antoniu)

Sources

HM-HIM Hadtörténelmi Levéltár (Budapest)
Polgári Demokratikus Forradalom iratai
Hadügyminisztérium 37. (légügyi) osztály
Magyar Tanácsköztársaság iratai
Hadügyi Népbiztosság 37. Légi ügyek osztálya
Keleti ill. Vörös Hadsereg Parancsnoksága, Hadműveleti iratok
Légügyi előadó
I-III. hadtestparancsnokság, Hadműveleti osztály és tüzérségi előadó
Légi- /repülő/ csapatok parancsnoksága
Honvédelmi Minisztérium iratai
37. (légügyi) osztály, 1919

Magyar Műszaki és Közlekedési Múzeum (Budapest)
Archívum
Kézirattár
303. Troján Gyula visszaemlékezései repülőtevékenységére.
Vojenský ústřední archiv - Vojenský historický archiv (Praha – via Jan Hammer, Jan Kaše és Petr Lukeš)
Fond Letecký sbor
1. polní letecká setnina

Romanian accident reports 1919–1921 courtesy of Dan Antoniu

Bibliography

AVRAM, Valeriu: *Escadrile româneşti deasupra Ungariei (1919).* - In. Aeronautica, 1993. Nr. 4. - p. 26–30.

AVRAM, Valeriu: *Războiul aerian deasupra României 1916–1919.* Bucureşti, Editura Militară, 2017.

AVRAM, Valeriu: *Aeronautica română în serviciul Marii Uniri.* Bucureşti, Editura Militară, 2018.

BARKÓCZY-KLOPSCH Béla: *Mackensen vezértábornagy hadseregének átvonulása Magyarországon az összeomlás után.* - In. Magyar Katonai Közlöny, 11. évf. 1923. - p. 188–230.

BLUME, August G. – CASIRATI, Alberto: *The Italian Campaign in Macedonia and Albania. Part I.* - In. Over the Front, Vol. 7/1. 1992. - p. 3–23.

BLUME, August G. – CASIRATI, Alberto: *The Italian Campaign in Macedonia and Albania. Part II.* - In. Over the Front, Vol. 7/2. 1992. - p. 111–129.

CIGLIĆ, Boris: *Krila Srbije. Wings of Serbia.* Beograd, Infinitas, 2009.

CZIRÓK, Zoltán: *A katonai repülés kezdetei Debrecenben (1918–1919).* - In. A Hajdú-Bihar Megyei Levéltár Évkönyve 35. (2018) - p. 61–72.

CZIRÓK, Zoltán: *Katonai repülők Mátyásföldön (1918–1919).* - In. Magyar Repüléstörténeti Társaság évkönyve, 2014. - p. 54–70.

CZIRÓK, Zoltán: *LVG C.VI felderítőgépek Magyarországon.* - In. Haditechnika, 48. évf. 5. sz. (2014) - p. 63–68.

CZIRÓK, Zoltán: *Hadirepülők Rákoson (1918–1919).* - In. Magyar Repüléstörténeti Társaság évkönyve, 2011. - p. 33–48.

CZIRÓK, Zoltán: *Az első légi háború Magyarország felett – 1919.* - In. Hadtörténelmi Közlemények, 124. évf. 2. sz. (2011) - p. 335–361.

CZIRÓK, Zoltán: *A Fokker D.VI-os vadászgép Magyarországon.* - In. Haditechnika, 45. évf. 5. sz. (2011) - p. 58–61.

CZIRÓK, Zoltán: *Repülőszázadok Ongán – 1919.* 1. rész: Ongai Kékdaru, 11. évf. 11. sz. (2009. november) - p. 10–11., 2. rész: Ongai Kékdaru, 11. évf. 12. sz. (2009. december) - p. 11., 3. rész: Ongai Kékdaru, 12. évf. 1. sz. (2010. január) - p. 10–11., 4. rész: Ongai Kékdaru, 12. évf. 2. sz. (2010. február) - p. 10–11.

CZIRÓK, Zoltán: *A magyar repülőcsapatok 1918–1919. évi történetéhez.* - In. Hadtörténelmi Közlemények, 122. évf. 3. sz. (2009) - p. 603–634.

CZIRÓK, Zoltán: *Német repülőgépek a Balkánról.* - In. A Repüléstörténeti Konferencia Közleményei, 2009. - p. 168–172.

FEDDERS, Peter A.: *The German Air Effort and the Central Powers Romanian Campaign Phase I, the Fluid Phase: August 1916 – January 1917.* - In. Over the Front, Vol. 14/1. 1999. - p. 25–35.

FRANKS, Norman – BAILEY, Frank – DUIVEN, Ric: *Casualties of the German Air Service 1914-20.* London, Grub Street Publishing, 1999.

GROSZ, Peter M. – HADDOW, George – SCHIEMER, Peter: *Austro-Hungarian Army Aircraft of World War One.* Mountain View, California, Flying Machines Press, 1993.

HERRIS, Jack: *Fokker Aircraft of WWI. Volume 6 – Foreign Service.* Reno, Aeronaut Books, 2022.

KAŠE, Jan: *1. a 2. polní letecká setnina.* - In. LK Speciál, Nr. 6. 2005. - p. 33–39. (Kaše, 2005/1.)

KAŠE, Jan: *Jediný Fokker D.VII (MÁG) v československém letectvu.* - In. Novinky MPM 26/05 - p. 3–6. (Kaše, 2005/5.)

KAŠE, Jan: *LVG C.VI.* - In. LK Speciál, Nr. 6. 2005. - p. 61. (Kaše, 2005/9.)

KAŠE, Jan: *Přílety chtěné i nechtěné.* - In. LK Speciál, Nr. 6. 2005. - p. 18–23. (Kaše, 2005/12.)

KAŠE, Jan – LUKEŠ, Petr: *1. polní letecká setnina.* - In. LK Speciál, Nr. 10. 2008. - p. 24–29. (Kaše–Lukeš, 2008/2.)

OWERS, Colin: *Fokker Aircraft of WWI. Volume 7 – Postwar.* Reno, Aeronaut Books, 2022.

PĂTRAȘCU, Gabriel: Unitățile de gardă ale Aeronauticii și Marinei Regale (1930-1947). - In. Revista Misiunea, No. 1/2020. - p. 233-246.

PETROVIĆ, Ognjan: *Vojni aeroplani kraljevine SHS/ Jugoslavije 1918–1930.* LET, Beograd, broj 2/2000.

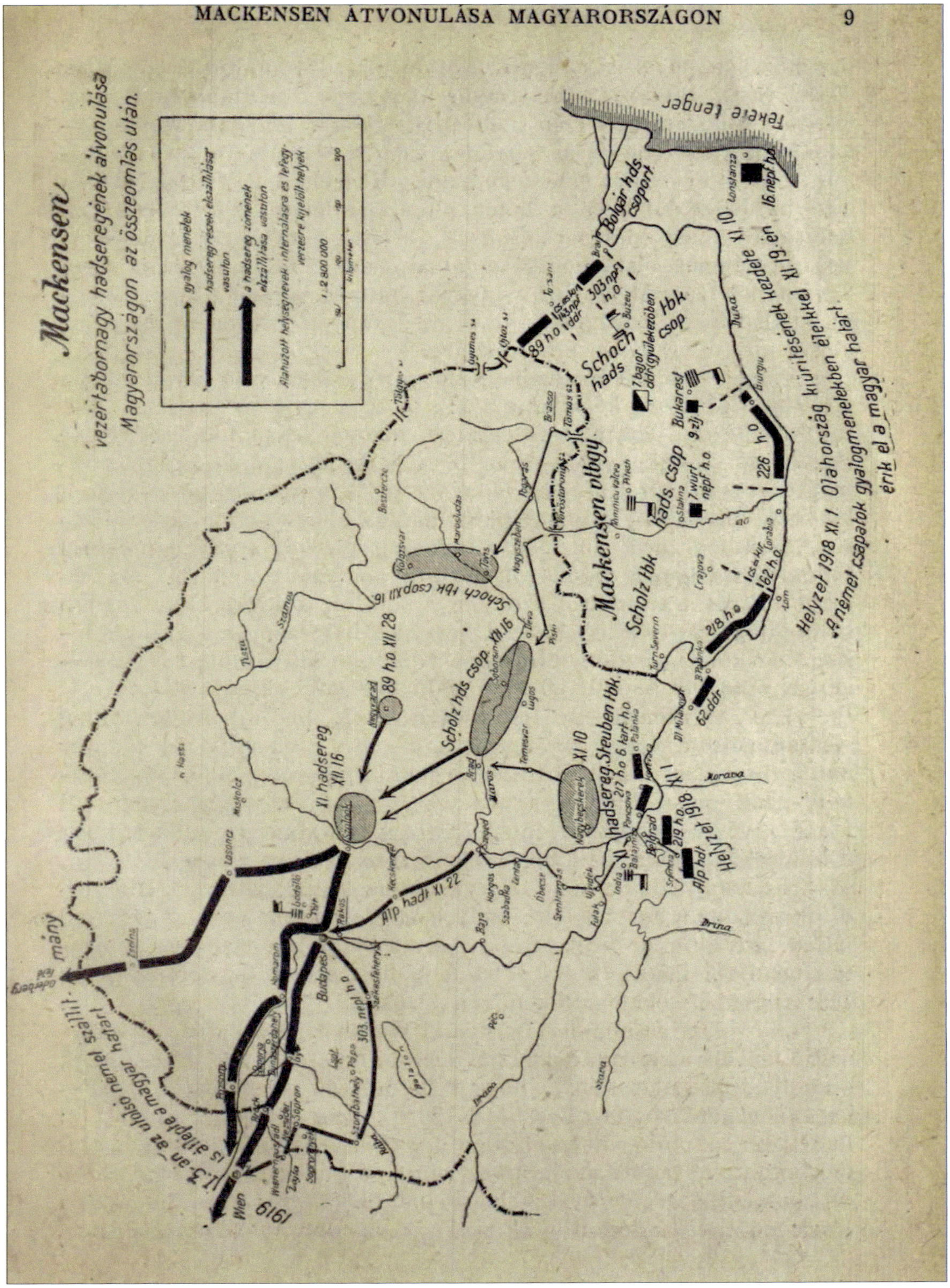

Above: Map sketch of the Mackensen army march through Hungary from a Hungarian study published in 1923. (Barkoczy-Klopsch)

Appendicies

Operational Air Units of the Hungarian Flying Troops from 23rd January 1919

I. Repülőcsoport parancsnokság (Budapest)
1. repülőosztály (Mátyásföld)
2. repülőosztály (Albertfalva)
3. repülőosztály (Győr)
4. repülőosztály (Kaposvár)
4. repülőosztály különítménye (Szombathely)
II. Repülőcsoport parancsnokság (Szeged)
5. repülőosztály (Szeged)
5. repülőosztály különítménye (Kecskemét)
6. repülőosztály (Szeged)
III. Repülőcsoport parancsnokság (Debrecen)
7. repülőosztály (Debrecen)
8. repülőosztály (Debrecen)

Repülőcsoport parancsnokság = Flying Group Command
repülőosztály = air flight
különítmény = detachment

Operational Air Units of the Hungarian Flying Troops from 24th March 1919

1. repülőosztály (Mátyásföld)
2. repülőosztály (Albertfalva)
3. repülőosztály (Győr)
4. repülőosztály (Kaposvár)
 4k repülőkülönítmény (Szombathely)
5. repülőosztály (Szeged)
 5k repülőkülönítmény (Kecskemét)
6. repülőosztály (Békéscsaba)
7. repülőosztály (Debrecen)
8. repülőosztály (Rákos)

Operational Air Units of the Hungarian Red Flying Troops from 8th April 1919

1. repülőszázad (Kaposvár)
2. repülőszázad (Albertfalva)
3. repülőszázad (Rákos)
4. repülőszázad (Győr)
 4k repülő különítmény (Szombathely)
5. repülőszázad (Kecskemét)
6. repülőszázad (Békéscsaba)
7. Székely repülőszázad (Debrecen)
8. harci repülőszázad (Mátyásföld)
Vízirepülő század (Csepel)

repülőszázad = squadron
Székelys are a Hungarian subgroup living mostly in the Székely Land in Romania
harci = fighting; here: fighter
vízirepülő = seaplane

Operational Air Units of the Hungarian Red Flying Troops from 28th May 1919

1. repülőszázad (Kaposvár)
2. repülőszázad (Albertfalva)
3. repülőszázad (Rákos)
4. repülőszázad (Győr)
5. repülőszázad (Kecskemét)
6. repülőszázad (Gödöllő)
7. repülőszázad (Mátyásföld)
8. harci repülőszázad (Mátyásföld)
9. vízirepülő század (Csepel)

Made in the USA
Monee, IL
05 November 2026

67948873R00038